Changing the game in home entertainment.

itisCloudTV™

User Experience Guide
(based on KODI 15.x)

itisCloudTV™ User Experience Guide
by Larry L. Broussard

Published on line at www.createspace.com. Printed in the United States.

ISBN-13: 978-1516991389
ISBN-10: 1516991389

This User Guide is the property of

Intelligent Technology Integration Solution, LLC
(IT IS, LLC)
5348 Crenshaw Blvd
Los Angeles, CA 90043-1810
1-323-348-4900

Additional copies may be obtained by calling the number above.

Contents

Figure 1: itisCloudTV

Introduction

Welcome to itisCloudTV™ and the future of home entertainment. Free TV is once again possible. If your have been frustrated with monthly cable or satellite TV bills, say hello to independence! Say "Hello" to itisCloudTV™!

itisCloudTV™ streams content from the Internet, which is quite different from traditional television programming. That's the uniqueness of the solution, because no longer must you wait for "them" to tell you what to watch and when to watch. With itisCloudTV™, daily time tables and programming schedules go away. Watch what you want to watch, when you want to watch it. This even includes movies that are still in theatres, as well as up-to-date episodes of all your favorite TV shows and series.

...making I T work!

Just about everything that has been produced for entertainment is available on the Internet - the "Cloud". The challenge has always been in harnessing the millions of links and trillions upon trillions of bytes of data that makes up the movies, music, videos, news, and other content on the cloud. itisCloudTV™ does.

itisCloudTV™ sets the trend for the future of television and home entertainment in the US, integrating proven technologies into an all-access capability, presented through a single device directly to your TV. itisCloudTV™ delivers a more complete entertainment experience, including in-theatre content. Finally, real budget relief from ridiculously priced cable and dish "programmed" entertainment is once again available.

What is Streaming Media?
Streaming is not new. It has been around for decades. Today, streaming media has become the less expensive alternative to traditional TV programming.

With streaming, movies, shows, music, most any type of rich media content, is delivered, packet by packet over a network, to a streaming device. The device "reconnects" the packets, and "plays" the rich media through your television or PC, as well as other viewing devices, including smart phones and tablets.

Many businesses are in the streaming media service (NetFlix, Hulu, Apple, Amazon, and many others), but comes with monthly fees. Internet "free" streaming, as an alternative solution, is very young in the US, because of the efforts of these and other companies, but it has been in existence for years. The rest of the world has streamed entertainment content, even from the U.S., for years.

Why Streaming Has Become Popular Today

In February 2015, the FCC passed the Internet neutrality rule, which basically states that traffic can't be charged an extra "fee" by Internet providers. This provision also applied to "air" traffic and WIFI networks. Technically, this means that anything on the Internet is free and can't be created or taxed by Internet providers. Service levels for such content must conform to typical Internet traffic -- no access or speed retardation due to its content type.

What this means for the public, is the simple fact that anything that can be found on the Internet is free. Many organizations non-profit and for-profit, have worked on streaming technologies for over a decade, and the newest and latest breakthroughs are making Internet streaming a viable option and the clear direction of the future in personal and home entertainment.

Cable companies and dish companies, as well as many "per-month program providers" don't want you to know what the rest of the world already knows: entertainment content that they sell to you on a monthly basis, is virtually free on the Internet. The Internet delivers content from around the world, even from the US. Others, and now you, won't pay a monthly fee for it.

So congratulations! You have joined the rest of the world on the information super highway for your home entertainment pleasures, as you save hundreds of dollars per month in the process. This will add up to literally thousands and thousands of dollars over your entertainment lifetime, and you're at the beginning of it all!

The plethora of options available to you through our itisCloudTV™ solution, speaks to the vast capability of the internet to deliver what cable dish and satellite companies don't want you to know -- that you can have a richer,

more extensive, and virtually limitless access to content for your entertainment pleasure, and you don't have to pay for it.

How does itisCloudTV™ work?

itisCloudTV™ connects to the Internet, so you must have Internet access. The ideal Internet strategy uses WIFI to connect, which enables you to place the device, and your television, wherever you desire.

Your Internet service provider can provide you with a WIFI hotspot; you can always call us for guidance in selecting a good service provider. Since this is the only monthly fee you pay, and since Internet access is all you need, it is best to select the best option you can get, based on your budget and Internet options available to you in your area.

In many parts of Los Angeles, CA, WIFI hotspots are provided to the public. While we have not explored this as an option, it is clear that the future of Internet access has a public focus, which will be supported in every major metropolitan area of the country.

If you are already able to access the Internet without having to pay, welcome BACK to the world of free home entertainment! Like it used to be! We envy you!

Also, get a wireless mouse and keyboard. It enhances your control of your device better than the remote. The AirMouse with built-in keyboard is the best choice.

How to Watch TV on itisCloudTV™

Everything you want to watch is on a stream, somewhere on the Internet. itisCloudTV™ helps you get to the stream you want to watch, and delivers your choices in a consistent, easy-to-understand format.

When you want to watch something on TV, simply connect to the stream and watch! That's it. No more missed shows or episodes. You can even watch a complete TV series from start to finish, or up to the present episodes available on the Internet.

As content is "played" on regular TV, it is also made available on the Internet. That means every episode is updated, and available for you to watch, almost as it "airs" on regular TV. The only difference is instead of waiting for "Saturday night at 11pm", simply pick your stream. This is especially true of re-runs, previous content, old-time favorites, and forgotten episodes.

The list of available entertainment on the Internet is virtually inexhaustible. Through itisCloudTV™, you can now experience the best of the Internet for free!

What to Expect

Technology that supports streaming media from the Internet is still relatively new. They continue to develop, so for now it is not perfect like traditional programming technologies. What you save in monthly fees have a tax on some quality, access, and speed, from the user experience side. You can expect "bumps" in the road to free home entertainment, as it exists today. This will change in the very near future.

Here's why: a streaming device does not connect directly to a streaming server on the Internet. Instead, the streaming device sends a request, and the server sends the media back in packets, millions of packets, one at a time.

Sometimes but not often, packets become scrambled, may not return and decode in proper order, or may simply not reach the device.

This does not occur all the time, just enough to possibly alarm the novice user. Not to worry. Whenever problems like this persist, simply unplug the device disconnect all the connections. Reconnect everything after five seconds. You will get a better connection.

Many of the issues you should expect to experience, are addressed in the Trouble Shooting section at the end of the Guide.

Just remember: every bump represents hundreds of dollars per year that stays in your pocket! So when you get disconnected from your entertainment choice, just smile, reconnect to your choice, and continue to enjoy!

The itisCloudTV™ Viewing Experience
When you turn on the device, the itisCloudTV™ Home screen is the first screen that you see. It is presented below.

Figure 2: itisCloudTV Home Screen

The itisCloudTV™ Home screen contains the familiar "Google search" field (at the top) and five icons (at the bottom).

The five icons on the Home screen are:

1. Kodi
2. Browser
3. Apps
4. Explorer
5. Settings

1. Kodi

Kodi is the application that makes streaming content from the Internet possible. It has been around for years, and is now up to version 15. (At the time of this writing the latest version is 15.1).

itisCloudTV™ uses the Kodi app to harness the Internet and render its content in an easy-to-navigate, user-friendly format. Whenever you want to watch TV, click the "K". (See the section entitled, "Inside Kodi").

2. Browser

itisCloudTV™ is much more than just another smart device for your TV. It is a fully capable Android computer. With it, you can surf the Internet right from your TV. Whenever you want to surf the Internet, click the "globe". You can also "Search" on the Internet by entering your search in the Google search field at the top of the screen.

3. Apps

itisCloudTV™ contains a myriad of Android apps, provided to help you make the most of your home entertainment experience. Many apps are already downloaded to your device. You can also add new apps to your device using Google Play. Whenever you want to run a different app, or find and install new apps, click the "circle".

4. Explorer

The Explorer app enables you to explore the contents of your device as needed. Although it is on the Home page, you would rarely need to use this app. When you need to copy data from your device to another (like a USB thumb drive or an external hard drive) click the "file cabinet".

5. Settings

The gears icon enables you to go into the Settings for your device. Use this app to configure your network connection, add a Google or email account, clear the cache (more on this later), and adjust other settings on your device. Simply click the "gear".

The next sections of this Guide look at Kodi version 15, as it is configured in itisCloudTV™. It presents many relevant insights that you will need to help you quickly learn how to use your device. Take time to read through the Guide so that it can help to make your TV experience as rewarding as possible, and so that you can get the most out of your itisCloudTV™ device.

You will save a lot of money with itisCloudTV™. It is to your advantage to understand how it works, as much as you can. This Guide will help you.

Inside Kodi

When you click the "K" , you enter the Kodi app. The first screen that you see (following initialization) is depicted below:

Figure 3: Kodi 15.x

There are seven (7) options inside Kodi:

1. VIDEO
2. SYSTEM
3. POWER
4. FAVORITES
5. WEATHER
6. MUSIC
7. LIVE TV

Let's look into each.

1. VIDEOS

Kodi starts out with VIDEOS as the first option.

Figure 4: Kodi VIDEOS Screen

Most of your content is delivered through this option, including some live TV, on-demand videos and movies, and content from around the world. (Additional live TV content is available through the LIVE TV option.)

The sub-menus under VIDEOS are:

    1. ADD-ONS - This option is the main option for accessing streaming content on the Internet. Movies, TV shows, podcasts, music, and even karaoke are available through the ADD-ON sub-menu of VIDEOS.

    2. FILES - You should never need to use this option, unless you have content of your own that you have uploaded to the device, or connected via an external hard drive.

    3. PLAYLISTS - This option gives you direct access to playlists that you create.

4. LIBRARY - This option gives you direct access to your personal video library. With an external hard drive (separate purchase), you can begin to create your own library of movies, music, videos and other content.

ADD-ON Apps in VIDEO

To watch a movie, TV show, or other content do the following:

1. Click the Add-on app of your choice
2. Select the type of choice (movies, tv shows, or what is provided in the list)
3. Select further options, such as gender, featured, in-theatres, favorites, etc., based on the choice you have within the app you use.
4. Select a stream (if applicable).
5. Enjoy the show.

The Apps Inside ADD-ONS

Several video apps are already provided in ADD-ONS.

Figure 5: Sample VIDEO Add-ons

Your list may vary from the picture above.

You can add and remove add-ons, based on your taste. (See "Get More" later in this section.

The main video add-ons to use for the best Internet streaming experience, including live TV, in-theatre, all-time favorites, children, adult, religious, PPV, etc., are identified in the next pages.

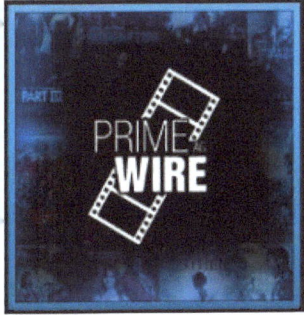

1. 1Channel/PrimeWire

1Channel/PrimeWire is one of the more stable of all the VIDEO apps available on the Internet. You can access movies, TV shows, personalized playlists, and other help options through this app.

When you enter PrimeWire, this is what you see:

17

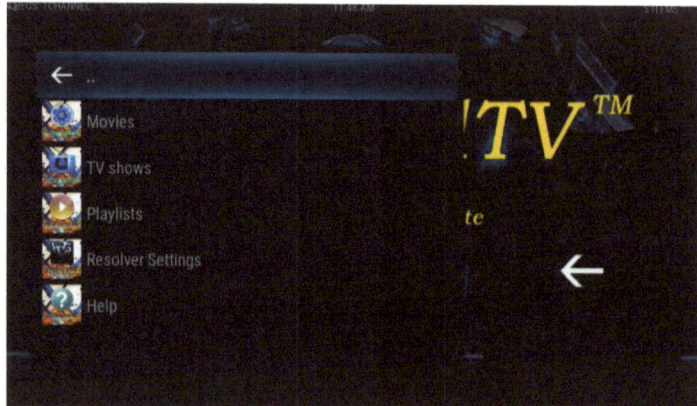

Figure 6: Inside 1Channel/PrimeWire

Make your selection from the options provided: movies, TV shows, Playlists, etc.

2. channel PEAR

Channel PEAR provides the greatest assortment of live streaming content.

It includes all the stations that people are typically used to seeing on TV, including major TV networks, premium stations (HBO, BET, TNT, CBS, NBC, etc).

18

When you click on channel PEAR, you will see the following:

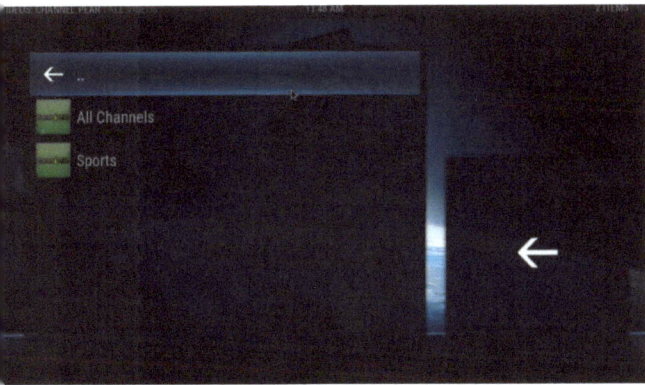

Figure 7: Inside channel PEAR

All Channels contain live streaming TV channels, and includes the channels that are listed under Sports.

There are over 200 live TV channels available in channel PEAR. The list is difficult to contain because channels are added and removed every day. All of the premium channels work fine, including HBO, BET, ION, TBS, TNT, ESPN CineMAX, ShowTime, NBC, ABC, CBS, and others.

Channel PEAR also streams international TV channels.

Use All Channels to find just about any live TV channel you want to watch.

3. GENESIS

Genesis is the senior statesman of streaming program delivery and is a fully stable app. When you enter GENESIS, you see the following:

Figure 8: Inside GENESIS

The GENESIS options include movies, TV shows, channels, Playlists, etc. Select the option you desire.

New movies and in-theatre movies can be found under MOVIES. Select IN-THEATRE for a list of available in-theatre movies available in GENESIS.

20

4. Phoenix

Phoenix is another very stable streaming app. It contains several internal apps that gives you a variety of choices for your entertainment pleasure.

When you go into Phoenix, you see the following:

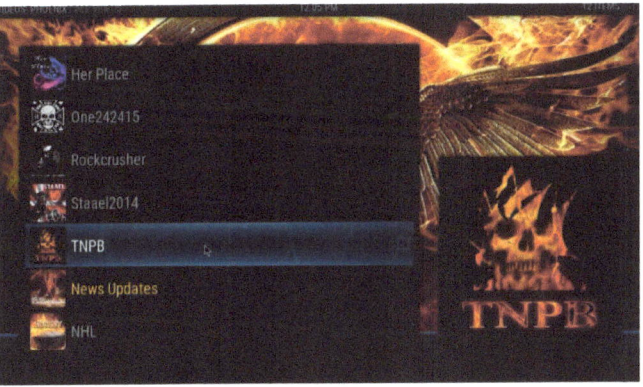

There are several apps inside Phoenix for access to a variety of content. Be sure to explore each app to understand what they provide for your entertainment pleasure. All of these apps and their content are free.

21

5. Get More

Get More is perhaps the most important icon in the entire set, because it enables you to add more add-ons to your VIDEO Add-ons list.

Click Get More to see the list of additional ADD-ONS you may add to your device.

This includes direct links to specific streaming content (such as Oxygen, CBS, NFL, NHL, adult streams, etc.

NOTE: INSTALLING OR ENABLING AN ADD-ON DOES NOT GUARANTEE THAT IT WILL WORK. CALL US FOR GUIDANCE BEFORE YOU CHOOSE TO ADD NEW APPS!

2. SYSTEM

The next option is the SYSTEM option. Here is the SYSTEM screen.

The Kodi SYSTEM option gives you access to the configuration options associated with Kodi. Your device has already been configured with the best possible content rendering experience. We spend many hours of effort programming the SYSTEM to give you effortless viewing without the need to reconfigure the device.

Whenever you want to add components to Kodi, use this option. The sub-menus under SYSTEM are:

     1. ADD-ONS - This option allows you to add or remove add-ons, enable existing add-ons, and also re-configure your user settings. BE CAREFUL USING THIS OPTION. FOR ERROR-FREE CONFIGURATIONS, CALL TIS, LLC FOR GUIDANCE WHEN USING THIS OPTION.

2. SKIN SETTINGS - This option allows you to change your "skin" settings, the manner in which Kodi is displayed, what options are available, and how content is delivered to your TV. BE CAREFUL USING THIS OPTION. FOR ERROR-FREE CONFIGURATIONS, CALL ITIS, LLC FOR GUIDANCE WHEN USING THIS OPTION.

3. FILE MANAGER - This option is used to set up the external locations of streaming servers and apps associated with your Kodi configuration. Your optimal Kodi experience has already been configured, so there is no need to access this option. BE CAREFUL USING THIS OPTION. FOR ERROR-FREE CONFIGURATIONS, CALL ITIS, LLC FOR GUIDANCE WHEN USING THIS OPTION.

4. PROFILES - not used.

5. SYSTEM INFO - This option provides a summary of the current configuration for your device, including: total storage, CPU usage, memory usage, screen resolution, network configuration, video configuration, hardware, personal video recording (PVR) services, and system uptime.

3. POWER

The POWER option is the only option that is configured with a different screen. SYSTEM looks like this.

Figure 11: POWER Menu Option

The POWER option enables you to exit Kodi and return to the itisCloudTV™ Home screen. The single sub-menu option is QUIT, which enables you to quit.

When you click POWER (the main option, instead of QUIT the sub-menu option), the following options appear on the screen:

1. EXIT - Exit Kodi.
2. SETTINGS - Use this to access the advanced system settings.
3. FAVORITES - Use this to access the favorites that are set up on your device.

To add items to your FAVORITES option list, do the following:

25

1. Find your viewing selection, and instead of clicking once, click and hold the choice for two (2) seconds. A menu box pops up, similar to this

Figure 12: Adding to XBMC Favorites

If you "Add to XMBC Favorites" (the first option), your selection will be available among your FAVORITES menu choices, the next time you return to FAVORITES.

4. FAVORITES
The FAVORITES option looks like this.

Figure 13: FAVORITES Menu Option

FAVORITES enable you to access your favorite shows and movies using a single click.

To add items to your FAVORITES option list, do the following:

1. Find your viewing selection (from within VIDEOS)

2. Instead of clicking the choice once to play, click and hold the choice for two (2) seconds long.

An "inset" menu box pops up, similar to this:

Figure 14: Adding to XBMC Favorites

The options are:

1. Add to XMBC Favorites
2. Search Icefilms
3. Add/Remove from Favorites
4. Add/Remove from Library
5. Add to CouchPotatoe
6. Show Information
7. Refresh Metadata
8. Watch Trailer
9. Mark as Watched

If you click "Add to XMBC Favorites" (the first option), your selection will be available among your FAVORITES menu choices, the next time you return to FAVORITES.

5. WEATHER

The WEATHER option looks like this:

Figure 15: WEATHER Menu Option

It displays the current city and conditions in the status bar, along with the present temperature, and current date and time.

When you click on WEATHER, the following screen appears.

Figure 16: Extended Weather Forcast Screen

You can change the city using the Settings option (under POWER).

6. MUSIC

The MUSIC option looks like this:

To listen to music:

1. Select MUSIC
2. Select music add-ons
3. The following screen appears.

The add-ons you see may vary. Some add-ons may require an account.

7. LIVE TV

The LIVE TV option looks like this:

To view the channels under LIVE TV, simply click LIVE TV, or click TV CHANNELS.

The group of pre-configured channels is displayed, which looks something like this:

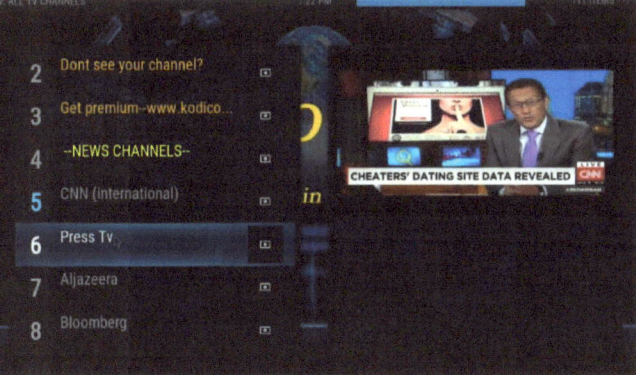

31

There are approximately 111 channels on the LIVE TV list at present; the list and the channels may change without notice.

SUMMARY - Inside Kodi

There is so much to the Internet and so much to Kodi. This tool gives you the ability to traverse the Internet for just about any type of information that is available through streaming media.

Education is the key to success and the Internet has it all. Enjoy more than a movie or your favorite TV show. Explore the rich world of information that is available to you from the Internet, thanks to itisCloudTV™.

Inside the Browser

If you are already familiar with the Internet, you already know the Browser. It looks like this:

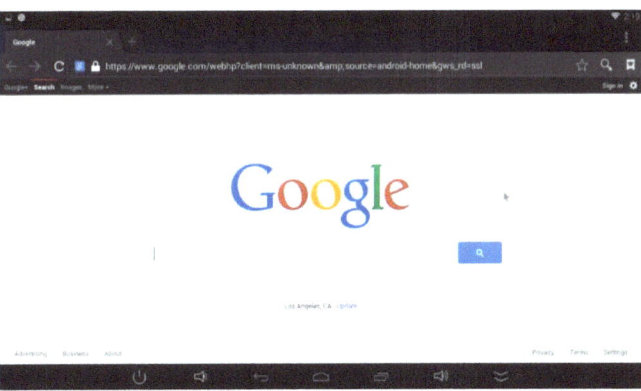

The Browser enables you to surf the Internet, just as you would on any other computer or tablet.

NOTE: When you use the Browser, you are not in Kodi, and none of the Kodi commands or screens apply to this app. Simply enter the url of the website you wish to surf to, and click.

Getting to the Apps

The APPS icon  takes you to the various apps that are already installed on your device.

When you click the APPS icon (above), the APPS window appears.

The APPS screen looks like this:

You can also add, change or remove apps from this screen.

The main app installer is Google Play. It looks like this:

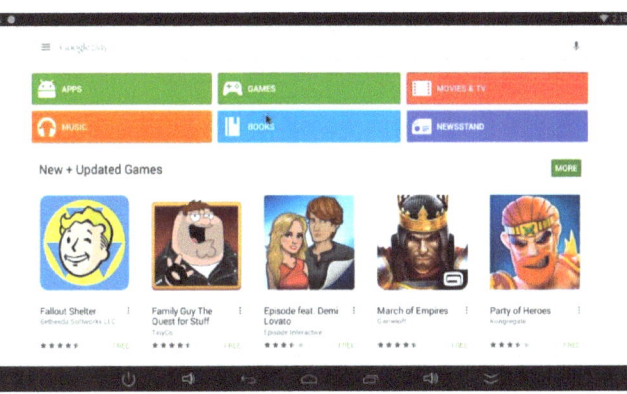

You need an account to access Google Play and to download additional apps from Google Play. The account is free, if you are not already a Google/Gmail user.

Many useful apps are already provided on your device. Explore what you already have; then decide if you need more, or less.

Explore Your Device

The EXPLORER icon 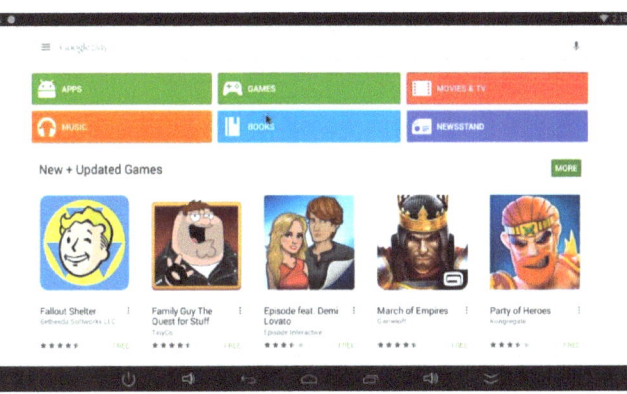 takes you into the file system of your device. It looks like this:

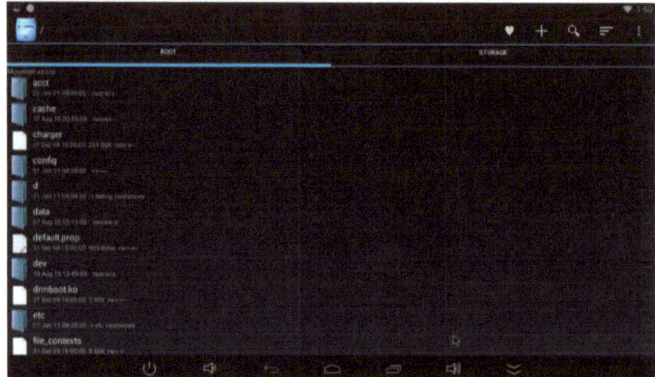

Be careful when entering the file system; many files that are necessary for smooth functioning requires the use of these files. DO NOT ERASE any files on your device that you did not put on!

Managing Your Settings

The SETTINGS icon gives you access to your device settings, including network, apps, storage, user account, and many others.

The screen looks like this when you first enter SETTINGS:

Adding your Network Connection

To add itisCloudTV™ to your WIFI network, do the following:

1. Click SETTINGS icon.
2. Make sure Wi-Fi (left hand side) is turned on. If not, turn it on.
3. Locate your Wi-Fi network on the right hand side.
4. Click the network.
5. Enter the password (as required)
6. If you connect, you will receive an IP address for your device. You will now be connected to your Wi-Fi network.

The Cache

The cache is the temporary data that your device maintains for immediate access to content you have looked at during the current session. It is added to or taken, based on your current choices and selections.

Cache data can become stale, and make your device act unseemly (crazy).

Often, it is necessary to clear your cache in Kodi. This can be done from SETTINGS.

To clear the KODI Cache:

1. Click SETTINGS.
   The SETTINGS window will open.

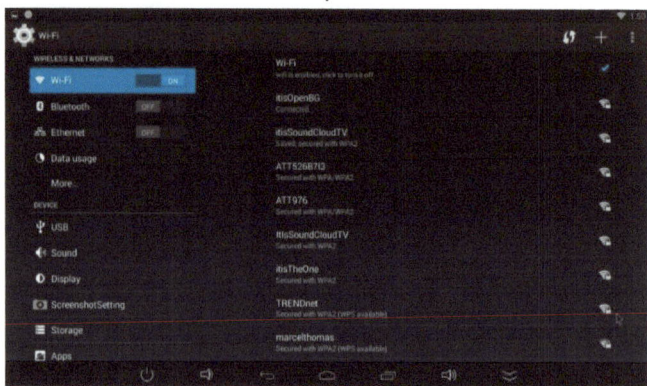

2. Click Apps (on the left hand side)
   The list of apps will display on the right hand side.

3. Click KODI (on the right hand side)
   The Kodi APP details screen will appear.

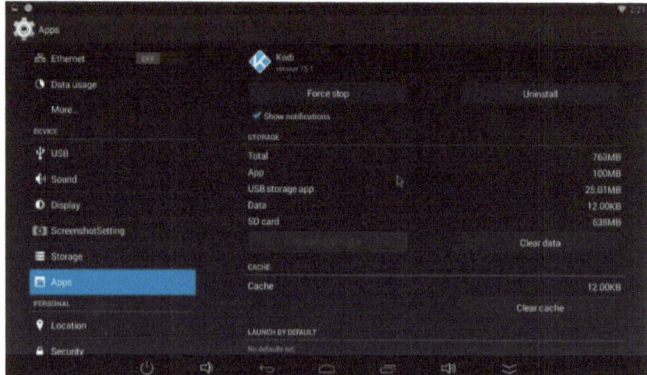

4. Click Clear Cache.
    The cache will return to size of 12KB.

That's it. Return to Kodi and enjoy.

Trouble Shooting Tips

On again, it is important to remind ourselves that streaming technology is continuously evolving. Here are some things you may experience on your itisCloudTV™ device and ways to correct the problem.

A.   *Device seems stuck. Does not go back when pressing the back button or the right mouse button.*

1. Move your mouse pointer to the top of the page.
    Drag the top down. You should see the HOME icon at the bottom center of the screen.

2. Click the HOME icon. The itisCloudTV™ HOME screen appears.

3. Click on "K" for Kodi. If the app reloads everything, that's okay. If the app returns you to where you were before, simply continue from where you were. The back button will work now.

B.   *Device buffers too much during playback.*

1. Buffering occurs when the streaming server or the Internet does not keep up with your device to deliver your choice. Usually there are two options:

    a. Pick another stream from the existing app. There are scores of streams for every selection, right within the app.

    b. Pick another app to render your selection. Apps typically deliver the same content but from different sources. Some sources are better than others.

2. Whenever buffering occurs, remember: the device is doing what it is suppose to: buffer the content until it is ready to present. Find another stream, or use a different app and buffering will stop.

C.   *There is No Display*

1. Unplug all the cables and attachments.
2. Hold the ON/OFF down for 10 seconds.
3. Reconnect all attachments.
4. Turn the device back on.

D. *Kodi resets the device during playback (returns to Home screen).*

This means that the cache is full. Simply clear the cache and start over.

To clear the KODI Cache:

1. Click SETTINGS.
   The SETTINGS window will open.

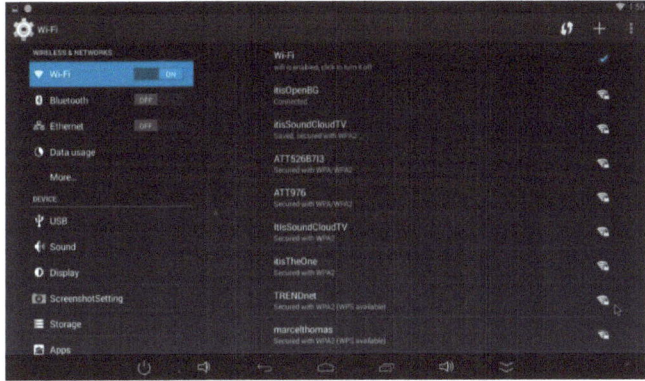

2. Click Apps (on the left hand side)
   The list of apps will display on the right hand side.
3. Click KODI (on the right hand side)
   The Kodi APP details screen will appear.

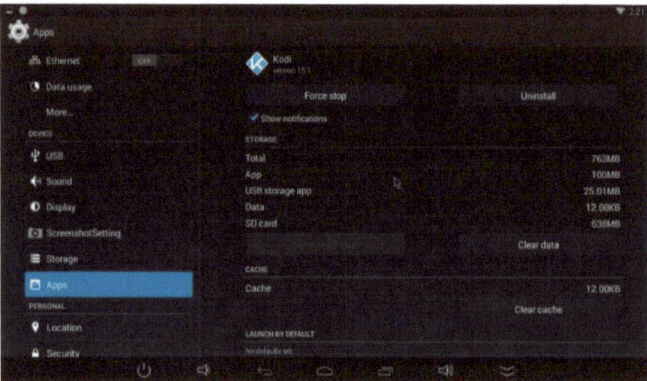

4. Click Clear Cache.

    The cache will return to size of 12KB.